Insects

by Kari Schuetz

BELLWETHER MEDIA • MINNEAPOLIS, MN

Note to Librarians, Teachers, and Parents:

Blastoff! Readers are carefully developed by literacy experts and combine standards-based content with developmentally appropriate text.

Level 1 provides the most support through repetition of high-frequency words, light text, predictable sentence patterns, and strong visual support.

Level 2 offers early readers a bit more challenge through varied simple sentences, increased text load, and less repetition of high-frequency words.

Level 3 advances early-fluent readers toward fluency through increased text and concept load, less reliance on visuals, longer sentences, and more literary language.

Level 4 builds reading stamina by providing more text per page, increased use of punctuation, greater variation in sentence patterns, and increasingly challenging vocabulary.

Level 5 encourages children to move from "learning to read" to "reading to learn" by providing even more text, varied writing styles, and less familiar topics.

Whichever book is right for your reader, Blastoff! Readers are the perfect books to build confidence and encourage a love of reading that will last a lifetime!

This edition first published in 2013 by Bellwether Media, Inc.

No part of this publication may be reproduced in whole or in part without written permission of the publisher. For information regarding permission, write to Bellwether Media, Inc., Attention: Permissions Department, 5357 Penn Avenue South, Minneapolis, MN 55419.

Library of Congress Cataloging-in-Publication Data
Schuetz, Kari.
 Insects / by Kari Schuetz.
   p. cm. – (Blastoff! readers: animal classes)
 Includes bibliographical references and index.
 Summary: "Simple text and full-color photography introduce beginning readers to insects. Developed by literacy experts for students in kindergarten through third grade"–Provided by publisher.
 ISBN 978-1-60014-774-6 (hardcover : alk. paper)
 1. Insects–Juvenile literature. I. Title.
 QL467.2.S37 2013
 595.7–dc23   2012000964

Table of Contents

All animals belong to the animal kingdom.

They are divided into many smaller groups of related animals.

What Are Insects?

Insects are one of the three
main **classes** of **arthropods**.
Arthropods do not have backbones.
They are **invertebrates** with
exoskeletons.

The Animal Kingdom

vertebrates

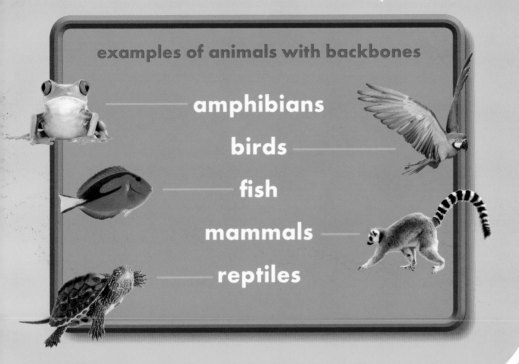

examples of animals with backbones

amphibians

birds

fish

mammals

reptiles

invertebrates

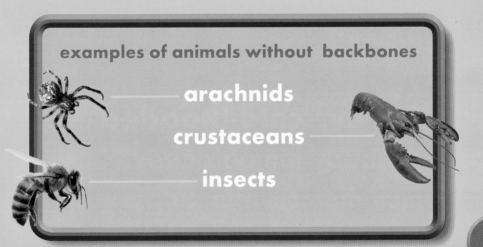

examples of animals without backbones

arachnids

crustaceans

insects

Insects are **cold-blooded**. Their body temperatures match their surroundings.

Some insects huddle together
to stay warm in cold weather.
Others become **dormant**.

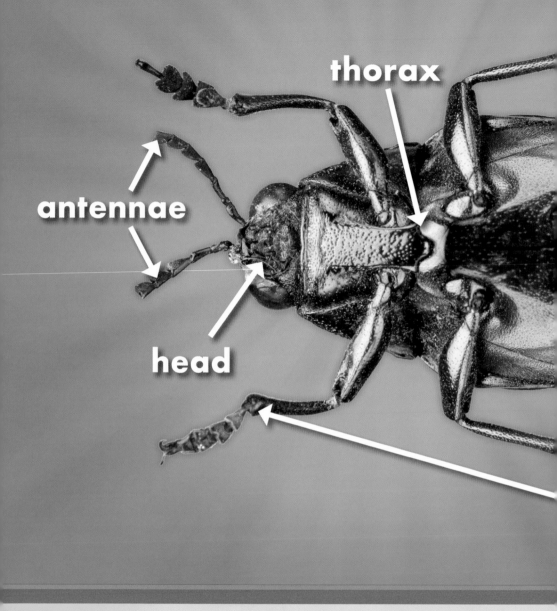

thorax

antennae

head

An insect's body is divided into three parts. They are the head, thorax, and abdomen.

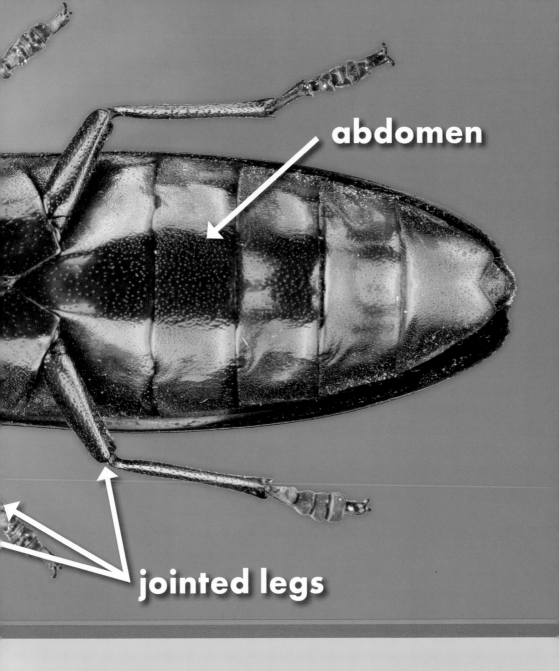

abdomen

jointed legs

The head includes eyes and
two **antennae**. Six **jointed** legs
are attached to the thorax. The
abdomen contains the stomach.

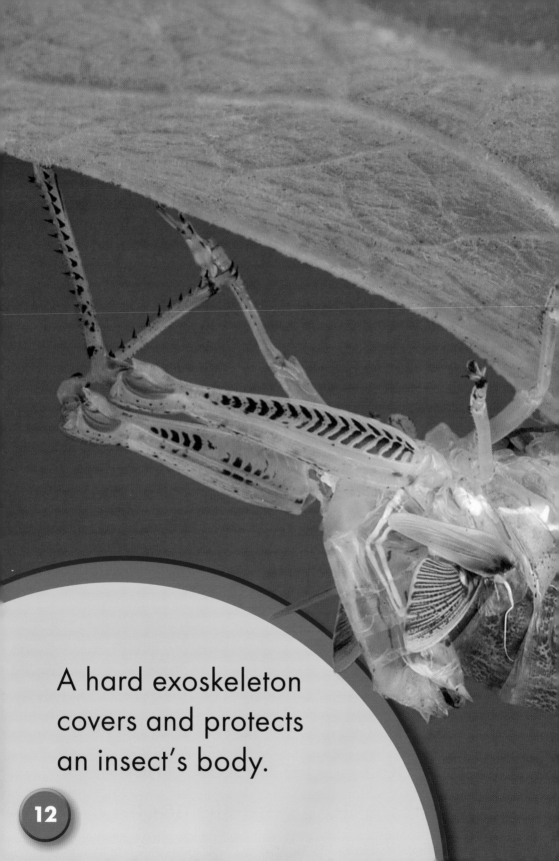

A hard exoskeleton covers and protects an insect's body.

Throughout life, insects **shed** their exoskeletons and grow new ones.

Most insects have wings. They are the only arthropods that can fly.

Flight allows monarch butterflies to **migrate**. It also lets bees move **pollen** between plants.

Insect Life Cycle

larva

pupa

All insects have at least three stages in their life cycle. These are egg, **larva**, and adult.

adult

Some insects also have a
pupa stage. A pupa goes
into hiding as it changes from
a larva to an adult.

Small Size, Big Impact

stick insect

Most insects are tiny.
However, some stick insects
grow to be 14 inches
(36 centimeters) long.

All insects are big when it comes to the **food chain**. They are a major source of food for many other animals and sometimes one another!

Incredible Insects

Largest:
Acteaeom beetle

Fastest Flying:
Green Darner dragonfly

Heaviest:
Goliath beetle

Loudest:
African cicada

Longest Life Span:
African mound-building termite

Shortest Life Span:
mayfly

Goliath beetle

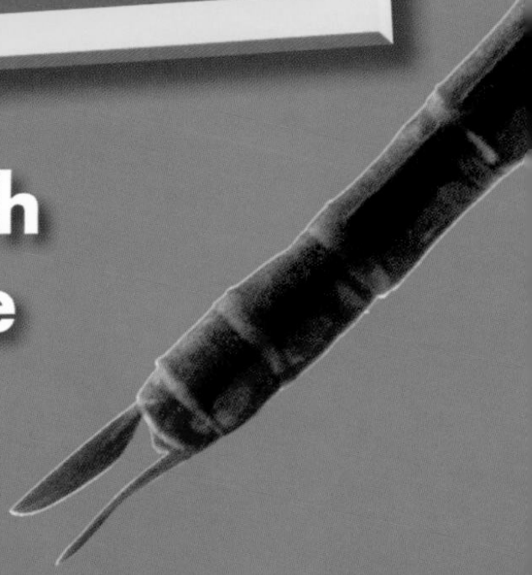

Green Darner
dragonfly

21

Glossary

antennae—organs attached to the head of an insect; antennae are used for smelling, hearing, tasting, and feeling.

arthropods—invertebrates that have divided bodies and three or more pairs of jointed legs

classes—groups within the animal kingdom; members of a specific class share many of the same characteristics.

cold-blooded—having a body temperature that changes to match the temperature of its surroundings

dormant—inactive

exoskeletons—hard protective coverings on the outside of insect bodies

food chain—a chain that shows what eats what in the animal kingdom

invertebrates—members of the animal kingdom that do not have backbones

jointed—having joints that allow for bending

larva—a wormlike baby insect that hatches from an egg

migrate—to move from place to place, often with the seasons

pollen—a powder in plants; insects help create plant seeds when they spread pollen between plants.

pupa stage—the third stage of life for some insects

shed—to let fall off; insects shed their exoskeletons as they grow.

To Learn More

AT THE LIBRARY
Bonotaux, Gilles. *Dirty Rotten Bugs? Arthropods Unite to Tell Their Side of the Story.* Minnetonka, Minn.: Two-Can, 2007.

Gray, Susan H. *The Life Cycle of Insects.* Chicago, Ill.: Heinemann Library, 2012.

Soloff-Levy, Barbara. *How to Draw Insects.* Mineola, N.Y.: Dover Publications, 2009.

ON THE WEB
Learning more about insects is as easy as 1, 2, 3.

1. Go to www.factsurfer.com.

2. Enter "insects" into the search box.

3. Click the "Surf" button and you will see a list of related Web sites.

With factsurfer.com, finding more information is just a click away.

Index